Ann Hamilton

corpus

Essay by Lawrence Raab

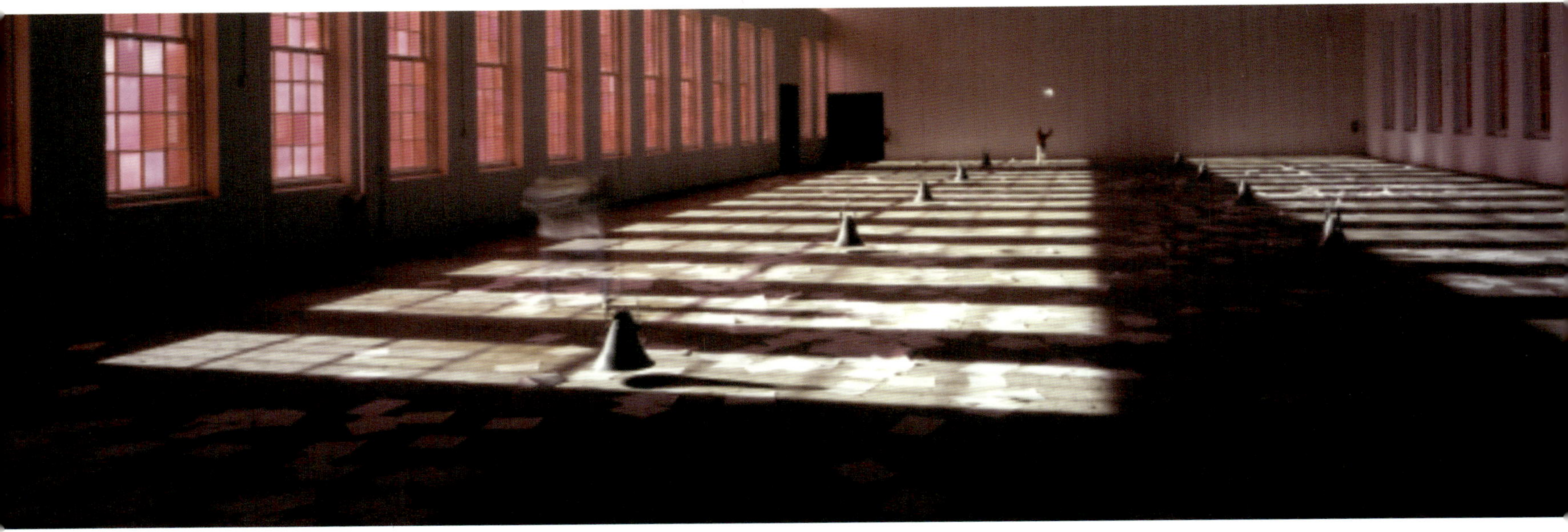

Acts of Finding

~ 1 ~

At first the gallery looks empty. Pieces of paper, scattered about on the gray concrete floor, haven't yet been swept up. What is there to look at? Could you have come to the wrong place? Then a single sheet of paper floats down from the ceiling. Then, far across the room's cavernous space, another.

What is here appears slowly. Emptiness, light, small things happening — these are the most immediate impressions of Ann Hamilton's *corpus*. More definable elements include: onionskin paper, silk organza, horn speakers, spinning speakers, rotating video, wood benches, and pneumatic machines. These machines — forty paper-handling mechanisms — are visible two stories up in the rafters of MASS MoCA's Building 5. Audible as well: a whoosh of suction, then a snap, and a page is released — here, there, in no discernible pattern.

As you walk into the gallery, blurred voices surround you, sometimes individual, sometimes communal, overlapping and echoing, then broken by silence — a recitation, a catechism. Twenty-four large, cone-shaped speakers have been suspended in two rows from the ceiling. Slowly and in unison they descend to within inches of the floor, pause, then rise. As they descend, the speakers form a kind of corridor through the space; as they rise, this passage slowly vanishes into the undivided air. At first, however, you may not notice that the speakers are moving.

Then you hear music beyond the voices — chanting, apparently — from the far end of the gallery. Then — but "then" doesn't seem like the right word, the way it so clearly organizes such disparate perceptions. Probably the light was what was most immediately striking, since each pane of each tall window on either side of the football-field-size gallery has been covered with crimson silk — 120 windows, 3500 individual panes. But the color of the filtered light on the white walls takes a while to register as an effect, just as the mostly empty space only gradually asserts itself as a presence. And the difficulty of organizing in time the various elements of the installation is shadowed by your initial impression that nothing, really, was there at all.

As you walk further among the speakers, certain phrases separate themselves from the drone of voices: "a word a body everybody everywhere..." Is there something specific that needs to be understood? Is there a clue? "...a unified song now lost..."

If there are clues, is there a solution? "...body after body place after place..." But if one word is no more or less important than another, one phrase no more or less revelatory, perhaps the idea of a clue isn't helpful. "Then," after all, did seem to impose the sequential too rigorously. "...hold fast to the unseen..." Certainly that sentence sounds suggestive. But what if you hadn't happened to hear it, if you'd let it all remain a wash of sound? Would you have missed something essential? Is there anything like a question to be answered, like a meaning to be uncovered?

~ 2 ~

"To equate my painting with symbolism," René Magritte wrote, "conscious or unconscious, is to ignore its true nature." Magritte would have admitted that his pictures *use* symbols. It's the translation of image into symbol that he wants to resist. The "true nature" of the picture doesn't deny its engagement with the symbolic; it denies being wholly absorbed into the symbolic. Magritte continues:

> People who look for symbolic meanings fail to grasp the inherent poetry and mystery of the image. No doubt they sense this mystery, but they wish to get rid of it. They are afraid. By asking 'what does this mean?' they express a wish that everything be understandable. But if one does not reject the mystery, one has quite a different response. One asks other things.[1]

What other things? Magritte doesn't say, but following the pattern of his thought can be helpful. Paintings are provocations. People prefer not to be unsettled, uncertain, in doubt. They turn to meaning for security. Symbolism can provide that, insofar as the symbolic translates the vagaries of the image into the solidity of an idea. An image is ready to do this work, and can deliver the symbolic as efficiently as it can produce its own name. But the name — the valise which is called a valise, the clock which is called a clock — is merely what's been attached to an object so we can use it, carry things around in it, tell the time. Other names will fit as easily, as Magritte's painting from 1936, *The Key of Dreams*, suggests. This picture is divided into four equal parts, each containing a single image on a dark background with a caption. The images have the flat realism of a child's first reading book, a thing paired with the name by which it is known. So a valise is called "the valise," but next to it a pitcher is called "the bird." Above the pitcher a horse is labeled "the door," and to the right of the horse is a clock called "the wind."

Magritte no doubt knew the risk of his work, how easily its "true nature" could be overlooked, or avoided. "A thing which is present can be invisible," he said, "hidden by what it shows."[2] Taken for granted, it fades from sight. But freed from the context of the familiar, the thing reasserts itself. The tortoise-shell comb in *Per-*

sonal Values (1952), by appearing huge in its ordinary room, or ordinary in its miniature room, reclaims its exact shape, rescues itself from over-familiarity. Then "the inherent poetry and mystery of the image" become available. A nervous viewer "no doubt" senses this mystery, and may well have been made uncomfortable by it. Why? Because mystery resists translation into meaning, which is not to say that mystery is meaningless, but that it actively *resists* becoming something other than itself. When the detective solves the case that mystery is over, and we are left with a world of reasonable explanations. Any aura of the uncanny or impossible has been dispelled. Strangeness has become fact.

'What does it mean?' is one of the central questions that mystery, at its peril, provokes, just as the detective story asks, 'Who did it?' What's important is not to avoid asking 'What does it mean?' (which in any case would be impossible), but to avoid settling for it. We may wish that everything were understandable, but we know it isn't. So if we do not reject the mystery, we ask "other things," one of which might be: Why did I need to know what it meant? Or: What do I want it to mean? Or even: How can I live with what can't be understood?

"The power of thought," Magritte writes in a letter from 1959, "is demonstrated by unveiling or evoking the mystery in creatures that seem familiar to us, out of error or habit." [3] We all know how hard it is to look freshly at what we're accustomed to seeing every day. That mountain, that tree, that building, which a stranger might well find arresting, we pass without noticing, since mere acknowledgment is a kind of blindness. But if the tree falls, if the building burns down, absence renders them visible. Destruction and death create for a while the achingly palpable semblance of what was lost.

Yet Magritte suggests that this "unveiling or evoking," this rescuing of mystery from indifference, need not be the result of violence, but can be generated by "the power of thought." The painting is designed to provoke such energy, which can eventually unveil the things of the world. But the power of thought can be defeated by the substitution of symbolic meaning for mystery, the problem being not so much the limitations of the symbolic, but the reflexive act of substitution. A moment of disturbance capable of producing the engagement of thought is replaced by something comfortable and secure — unearned meaning, demystified symbolism, "facile self-assurance," which, according to Magritte, must always be countered by "appropriate mistrust." [4]

Ann Hamilton's *corpus* begins with emptiness, the sense of a void. Then we see things happening, mysterious events that work to provoke "the power of thought," that have been constructed to challenge and shape the viewer's thinking. "In making work that is visceral and experiential," Hamilton said in an interview in 1990, "what I'm trying to do is keep you at that place where the first impetus is not to name and rely on literary or psychoanalytic traditions. Certainly, all of that can inform the experience, but it can also get in the way. So much of our experience we give over to specialists who say, 'This means this, and so that's what it's about.' If there are any poetics in my work at all, it's really about resisting that..."[5]

Hamilton wants her installations to physically envelop the viewer before they allow an intellectual response. She hopes to "keep you at that place where the first impetus is not to name," but she knows we won't stay there very long. The impulse to interpret is too strong. But some "appropriate mistrust" can be introduced. The equation of "this means this" can be complicated. Hamilton's "poetics," like Magritte's, depends upon a drama of resistance, in which the viewer's predictable inclination toward meaning is acknowledged, but purposefully disrupted.

~ 3 ~

'What does it mean?' is a question *corpus* is prepared to ask in a variety of interesting ways, but only after the viewer is confronted by several other consequential questions, the first of which might be 'Where is it?' Initially *corpus* can appear invisible, or at least diffuse. But soon 'Where is it?' becomes 'What does it consist of?' If your first impression of *corpus* was that nothing was there, that would have been shortly after it opened, or after every two-week period, when all the paper is swept up and removed. But it doesn't take long for the paper to accumulate to the point where attention is immediately called to it; a little more time and there are drifts, like snow, through which the visitors wade, or in which they play. Children catch and collect the falling pieces. Or, as the piles grow, kick them around, fall into them, throw handfuls at their friends. Is this appropriate? How, you might wonder, are you supposed to act here?

Magritte's meditations on the power of the image apply to the flat unchanging surface that is a painting. You see it all at once, propped on the easel or hung on the wall. When you turn away the power of that image may have altered your sense of the world, but what you see is everything that the painting doesn't consist of. To re-experience that picture you know exactly where to look. In an installation, on the other hand, every part asserts some kind of claim. "You can't be inside a painting," Hamilton has said, "and I want people to be absorbed into a physical kind of mass."[6] While you are in *corpus* everything you see is attached to the work, including yourself. 'What does it consist of?' threatens to become 'What doesn't it consist of?'

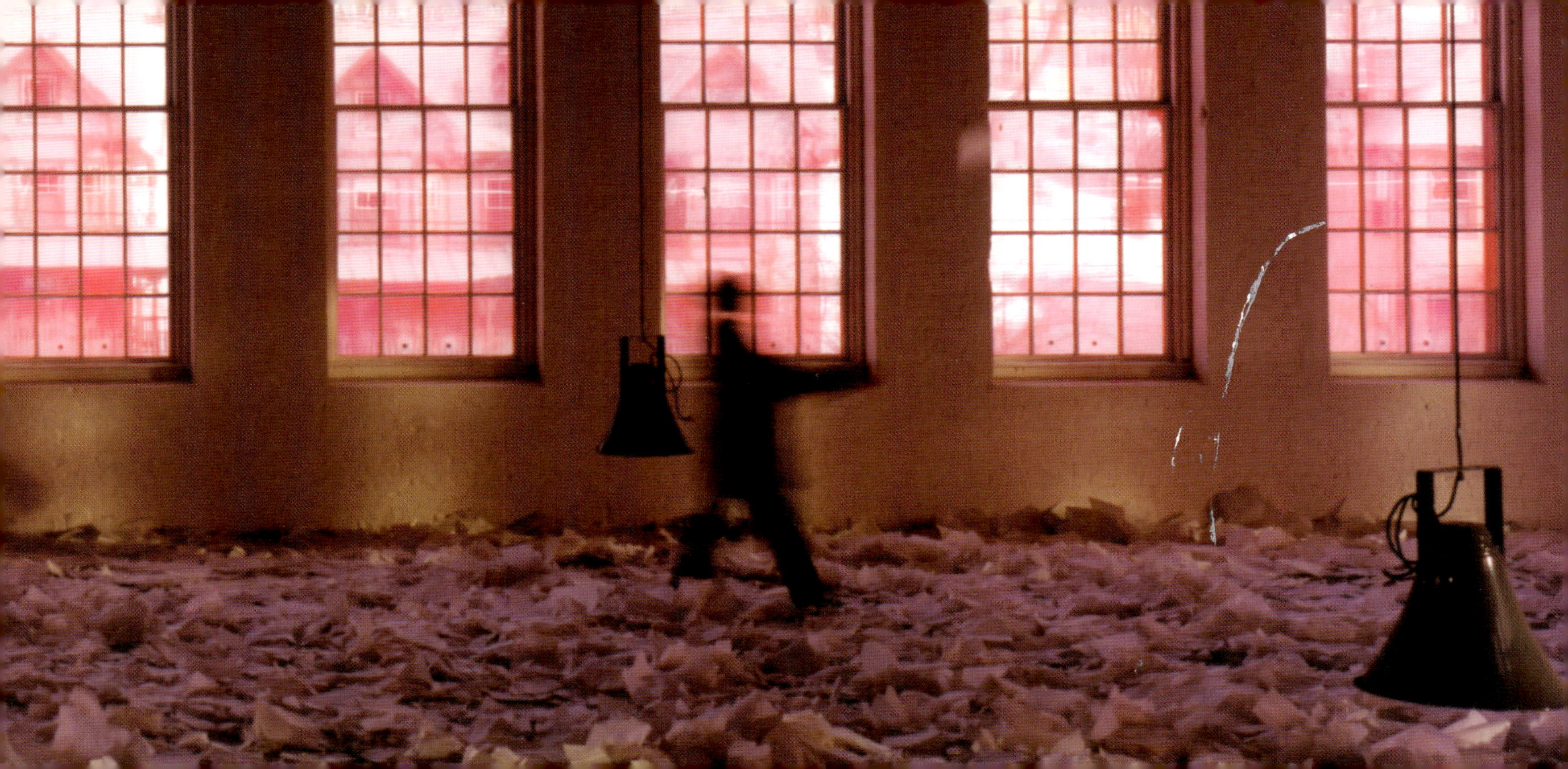

~ 4 ~

That *corpus* consists of three separate and quite different rooms isn't immediately apparent. But walking through the main gallery, under and around the falling and gathering paper, you hear music coming from a room beyond the one you're in. This second gallery is small and illuminated only by a single window of unfiltered light. The walls and floor have been painted dark gray, which shades into black in the corners, and the ceiling feels particularly low, given the more than two-storey high gallery you've just left. Four small speakers are spinning above your head, and not far above it either. The effect is unnerving, even menacing, and at odds with the serene quality of the wordless vocal music (a piece called "liquid air" by Meredith Monk, composed for a collaborative theater piece with Ann Hamilton entitled *mercy*). There's a sense that the speakers, which not only circle on extended arms but rotate individually, could descend. (And why not? You've just seen speakers descend.) But these, as they turn, can't be seen clearly, are always whirling away from your line of sight, a little like bats. Turning to follow the track of their movement is dizzying.

And now an unpleasant whine from the machinery, inaudible outside the room, becomes annoyingly apparent. Is this an effect, a way of making the space feel even more uninviting? Or, like the malfunctioning machine, is it something you should try to ignore, set aside as not part of the installation? But is anything not part of the installation? The act of separating what *corpus* is from what it isn't may be an impossible task, but comparisons are useful. In the second gallery the elements of the first have been reversed. Speed replaces slowness. Darkness replaces light. What was expansive becomes constricting, what was lyrical and playful becomes jittery and threatening. If the first gallery seemed like an appealing place to linger — a kind of town square if there are other people, or if you're alone, a calm, contemplative space — the second room is one most visitors will want to hurry through.

And yet — it is possible to experience the room quite differently. Or to imagine it differently, then experience what has been imagined. The music, after all, is alluring. Unburdened by language, therefore by all of the problems of meaning, it's the music of trance and meditation. And darkness — what is it *like*? The possibilities are multiple and contrary. The darknesses of fear, alienation, and ignorance can be as different from each other as they are from the darkness of comfort, or solitude, or peace. *Darkness* remains the same, even as the feelings projected onto it shift and change. Too easy, then, to claim the space of the main gallery as friendly and communal; surely it could be felt as unfocused, isolating, chaotic, overpowering. Space, like nature, has no inherent value.

If the whizzing speakers are like bats, they are also fixed in their movements, and therefore unlike bats. Standing for a while directly beneath one as it circles can even generate a sense of safety; the visible threat, stationed above, now becomes protective. Speakers as bats, bats as aggressors, then guardians — in fact we know that bats feel nothing toward us. They move to eat, eat to live. And perhaps, pausing in the dark room, you don't even imagine bats. The music urges you to stay, fall under its spell, like the sirens' song — irresistible, and deceptive. Or you don't think of Odysseus tied to his mast either, but of something different, something else entirely. What anything might be like is always changing, and the aim of *corpus* is to keep the viewer aware of the process of change — how the action of a body moving through space transforms, moment by moment, the meaning of that space, how light alters as it illuminates.

The only way to go on from the dark room is to climb the stairs that lead to the third space overlooking the main gallery. Here you can see where you were, and you may realize with a certain unease that as you strolled around down there you were being observed, as now you observe others, who are probably also unaware of being seen. Looking down — and even in a sense back — you may feel like you've come to the end of a journey, arrived at a prospect, an elevation, a kind of scenic vista, an appropriate place for contemplation.

This open-ended room, with white walls and a polished wooden floor, contains three rows of low white benches made from heavy wooden beams, thirty in all. In the tall windows on the two outside walls, the individual panes have again been covered with silk, but doubled, so the color is more intense — crimson, deep pink, rose, changing with the light. In contrast, the four panes of each bottom row are an opaque white. These small changes in the handling of the windows give the upper room an intriguing formality. It's *like* the main gallery but different enough to register as a variation, or an extension of a theme.

Hamilton's title, *corpus*, is defined by the gallery card as "a collection of writings; the body of a person or animal; a collection of utterances." The operative word is "body," and its suggestive double meaning — an assembled body of work, and the living body of a person. Human bodies (and quite a few of them, given the complexity of the project) created this work, which is a body of utterances, images, and effects, through which the body of the visitor moves. One body is dependent upon another. The play on words is a nice clue, since it raises the concern explored by Magritte's *Key of Dreams*: how a word can float from one thing to another, illuminating both similarity and difference, in the manner of a metaphor.

~ 5 ~

"Metaphor consists in giving the thing a name that belongs to something else," Aristotle asserts. As its origin in the Greek word for *transfer* suggests, metaphor is a process of conveying, as well as of sharing, some quality. Hamilton's *corpus* is an accumulation of likenesses, a construction of metaphors. Any part of it invites comparison with something else within it, even as any element can evoke a likeness outside it. We are accustomed to using likenesses, consciously or unconsciously, to communicate the appearance of a place or the effect of an experience, as well as to fix meaning, so that we can go about our lives and say, for example, 'Where is my valise?' or 'Please pass that pitcher of milk,' with some assurance that a particular intent will be conveyed. But as Magritte's *Key of Dreams* illustrates, that process can be disrupted. If the thing is given a different, seemingly inappropriate name, suspicion is created that implicates both word and image.

The clock with its caption of "the wind" looks like a metaphor, though it may be a metaphor that doesn't work, or works only to reveal difference instead of likeness. As art it's a playful piece of subversion, although play, for Magritte, is always consequential. If the phrase "the clock which is the wind," or "the clock of wind," were to appear in a poem, it might seem persuasively surreal, or perhaps merely fanciful. But how easily a sensible mind can transform the strangeness of "the clock of wind" into the numbingly ordinary "winds of time."

When a thing is too obviously like something else the metaphor fails to generate any visual or intellectual excitement. It only makes sense. It's the illustration we didn't need, bad pedagogy. But the further apart two things are the more dangerous and demanding the metaphor becomes, the more we're aware of the evocative difficulty of the act of transference, as in the always startling opening of T. S. Eliot's "Love Song of J. Alfred Prufrock":

> Let us go then, you and I,
> When the evening is spread out against the sky
> Like a patient etherised upon a table…

The "unveiling or evoking" that Magritte talks about when citing the "power of thought" is essentially a function of metaphor. But instead of making strangeness familiar, Magritte's metaphors summon forth mystery from the apparently ordinary, just as the third line of "Prufrock" completely destroys the pallid romantic expectations of the poem's first two lines. The predictable beauty of an evening that might be like some assemblage of angels or gossamer veils is suddenly blown apart, and there, drugged and spread out between reality and dream, is Prufrock's evening, violently reconfigured.

"A metaphor," Paul Valéry writes in *The Art of Poetry*, "is what happens when one looks in a certain way." One appeal of this definition is Valéry's suggestion of action – a metaphor is more than a certain kind of comparison, it is "what happens" because of a particular way of looking. It is a kind of collision of two things that are both similar and dissimilar, a joining that results in a momentary fusion of the two, but also an awareness of the necessary separateness of both.

Likeness is temporary. "All metaphor breaks down somewhere," Robert Frost writes in his essay, "Education by Poetry." "That is the beauty of it." Part of the beauty and importance of the metaphorical *process* lies in knowing how far you can go with a metaphor, when it will start to fail, and therefore when you can no longer trust it. For Frost this is much more than an issue of craft, since the way poetry uses metaphor "goes on to the profoundest thinking we are capable of." Poetry itself is a kind of metaphor, insofar as it is like all the important thinking we can attempt, a kind of template for the action of the mind. And so, "unless you have had your proper poetical education in the metaphor, you are not safe anywhere. Because you are not at ease with figurative values…" "Values," Frost says, not just "language."

In a poem love is like a rose until the moment the reader starts to imagine thorns. That's where the metaphor breaks down, and where the poem fails, unless the poet has managed to move on. Or perhaps the poet has decided to include that collapse as a tactic within the poem itself, a way of rescuing the rose from the indifference of the familiar through the introduction of pain. Both of these gestures — the tense relationship of similarity, and its ultimate breakdown — are essential elements in Ann Hamilton's *corpus*, which assembles and puts in motion various possibilities for metaphor-making. What are we reminded of by the light, the paper, the voices? What's any of it *like*? And where does it lead?

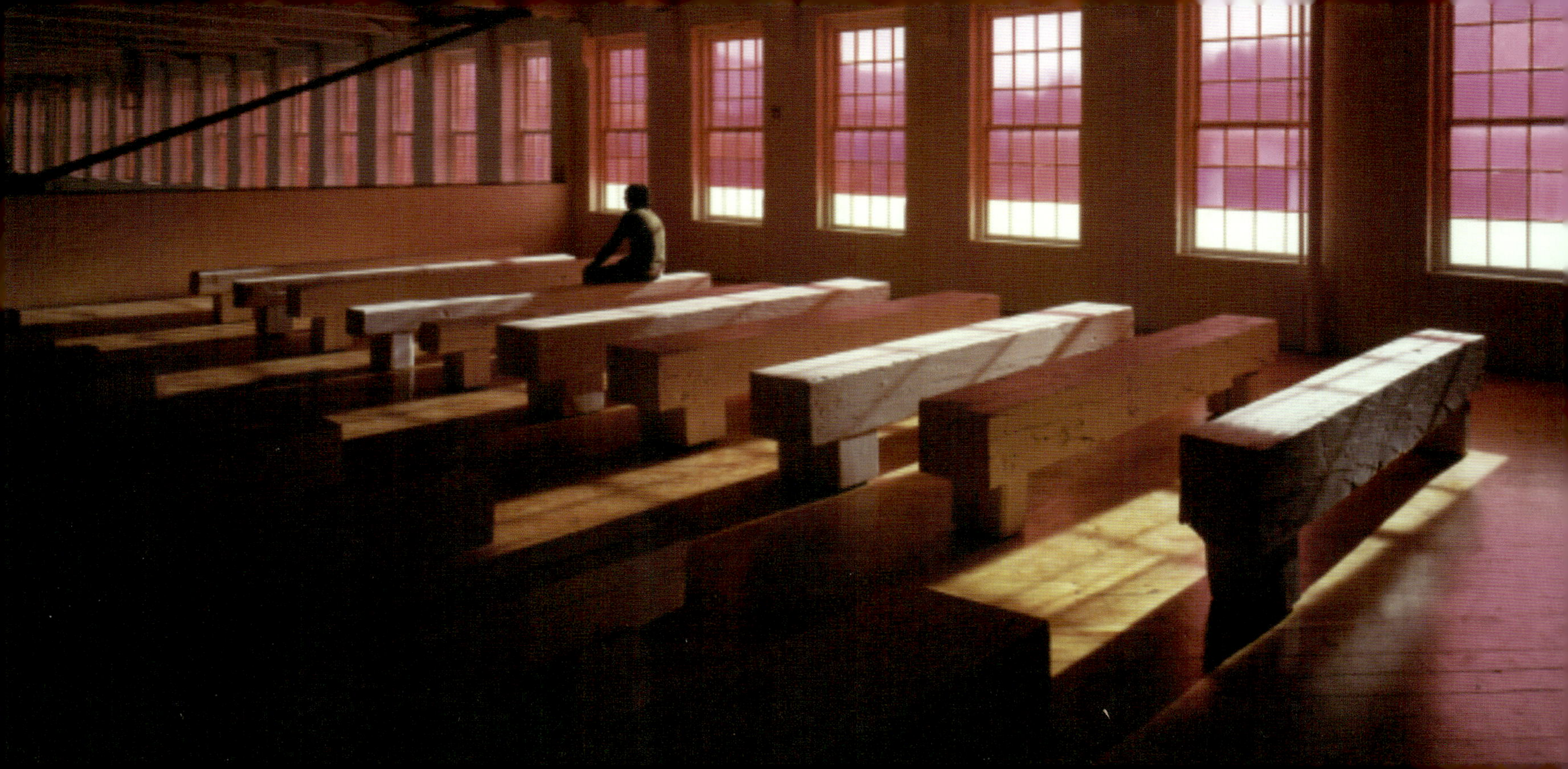

~ 6 ~

After you've walked up the stairs to the third gallery you may not be particularly tired, but since the room is full of benches, sitting down feels appropriate. If the benches were pews in a church, you would sit, then rise when the ceremony called for it, then sit again. If you were in an art museum, the most important rooms would probably contain benches so you could rest and study more carefully the masterpiece in front of you. If you were in a theater — as in a sense you are — you'd wait for the performance to begin. The benches in *corpus*, however, are worrisome. They make contemplation more difficult.

First of all, it's hard to know which way to sit. Do you face the main gallery, the apparent focus of attention? If so, a three-and-a-half-foot-high brick balcony prevents you from seeing into that space. Or do you look the other way, at a white wall, seemingly blank, but in fact a kind of screen across which projected words are sweeping? In a church — or in a theater, or an art gallery — you'd know which way to look, what you were supposed to see, and where the locus of authority was positioned — performer, minister, or masterpiece. The confusion Hamilton creates through the simplest of means — crude benches — subtly raises the issue of authority, then challenges it by dividing the attention that authority requires.

These benches, fashioned from beams that were removed from the old factory MASS MoCA once was, suggest the possibility of an assembly, a congregation. But no such group is likely to gather here. The benches of *corpus* are at odds with their metaphorical situation. They conjure up the image of a church, or a Quaker meeting room, then adjust and confuse it. This is Ann Hamilton's essential gesture: complication through suggestion. A memory is visually invoked, then set in a tension with other structural elements that make any automatic, sentimental, or unexamined response difficult. Remember, but worry about it, the room says. These white benches are like ghosts of a more easily defined past — a past we're always tempted to think was less vexed than our present. People gathered then, rituals were performed, someone in authority spoke, and the occasion ended. Here you sit and see the backs of other visitors looking down at what now is hidden from you — the vast room below, the people moving through it, that congregation to which you once belonged.

Or you turn the other way. Then you're confronted by a rectangle of light sliding across the wall, light in which there are words, although only a few letters are clearly visible at any moment. It seems that the letters are being individually typed out, but in fact the camera skips from one already formed letter to the next, suggesting the movement of a space bar. Following the projection as it circles the room is hard, since twisting around on the bench is awkward. Standing and turning creates an effect not unlike the dizzying attempt to follow the spinning speakers in the room below. In either case, Hamilton has made the situation difficult. She could have printed the words right on the wall. Instead, you have to work to see them, and concentrate to figure them out. This is what you end up with: "beginning," "the," "in," and "was," in various configurations. "In the beginning was the was beginning the in the beginning was beginning the beginning in the beginning was the was beginning…"

The difficulty of determining the text that *corpus* finally provides adds to its sense of importance. You had to figure it out. You've gone from the blank page to the writing on the wall. But these words are a kind of perpetual stammering of significance, a not-so-still point of inconclusive re-creation: Genesis, of course, and the beginning of the world: "In the beginning God created the heavens and the earth." But if remembering takes you in that direction, you could also go to the Book of John: "In the beginning was the Word, and the Word was with God, and the Word was God."

In each case the lines that follow are relevant and illuminating. From Genesis: "The earth was without form and void, and darkness was upon the face of the deep; and the Spirit of God was moving over the face of the waters. And God said, 'Let there be light'; and there was light. And God saw that the light was good; and God separated the light from the darkness." And from John: "He was in the beginning with God; all things were made through him and without him was not anything made that was made. In him was life, and the life was the light of men. The light shines in the darkness, and the darkness has not overcome it."

Both passages are about the act of making, and the difficult relationship between darkness and light. At the beginning of Genesis light emerges from the spoken word, and in an assertion of form upon the void, God separates it from the darkness. The thing becomes itself by being different. Before this the earth was "without form"; it was a void. But it was also "without form *and* void," since a void can only exist in contrast to a presence, as the darkness becomes darkness only when there is light to oppose it, and by opposing define and reveal it. God's act of separation reverses the process of making a metaphor, but creates those definably different entities that eventually will require the invention of metaphor.

At this point the symbolic equation of darkness with evil isn't firmly established, although it has been implied. God sees that the *light* is "good." In John the light becomes more determinedly and powerfully metaphorical: it shines within the darkness, even as the darkness attempts to "overcome it." Separation leads to opposition, which leads to conflict. The third room of *corpus* sends us back to the myth of origin, but leaves us spinning within it: "In the beginning was beginning."

The words keep circling — a text working to suggest revelatory meaning without defining it. Just as a desire for knowledge and closure is invoked, the words remain locked in the flux of a sentence that never ends. That these particular words return us to two different sources means we are made aware of a choice, but given no way to choose. Why should we want to choose? Because we desire certainty. Because we want to understand, as Eve yearned for the knowledge of good and evil, which the serpent promised would make her "like God." And because we are afraid, as Magritte says. Having fallen into knowledge, we continue to wish "that everything be understandable," the way everything must have been for our first parents, before they were deceived by desire.

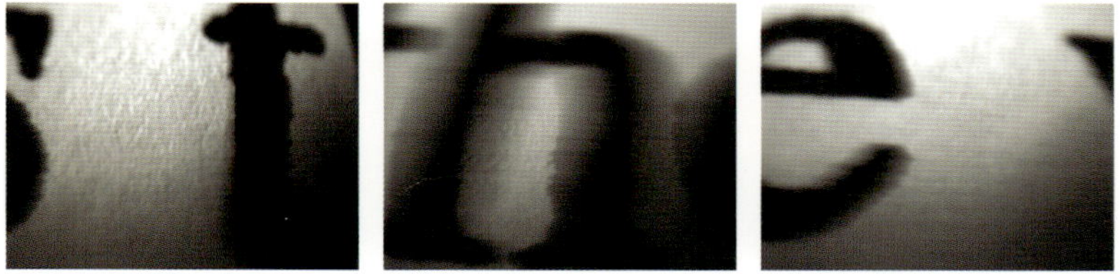

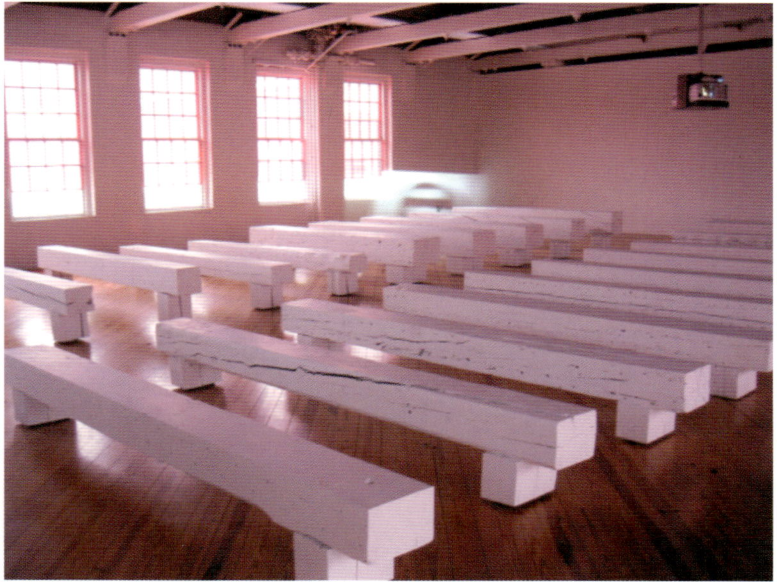

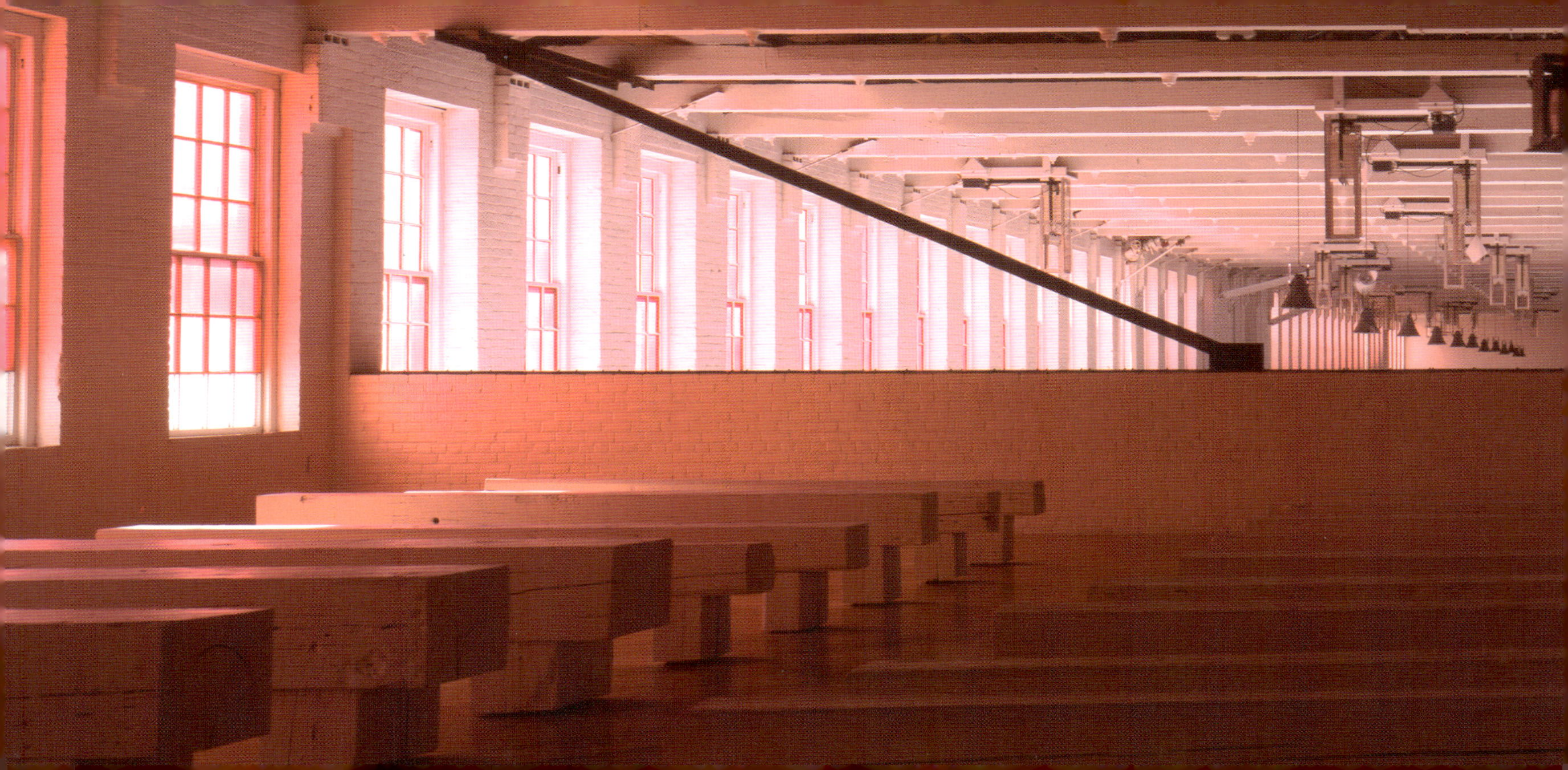

thought, refusing to become a mere screen upon which anything can be projected. The projections of language on the white walls are specific, fleeting, and endless. The installation itself is a series — a dramatic *sequence* — of metaphors that are always unstable but never incoherent.

Meaning adheres to them, but doesn't stick. Every metaphor breaks down somewhere. The work is designed to provoke and then resist symbolic meaning. This act of resistance leads to Magritte's "other questions." We can't not think. But we can say: Here's something else, something that complicates the matter. An account of this engagement becomes a way of describing the work itself.

"The meaning of the piece comes out of what it is physically," Hamilton told an interviewer before *corpus* opened. "It doesn't represent something else. It means what it is." [8] But Hamilton understands the alluring and unavoidable suggestiveness of metaphor, how anything can become *like* what it isn't. So one thing represents another, and stands in its place for a while, but is not absorbed into it. The rose is like love but is still a rose. The benches are like pews. That space is like a cave, or hell. This space is like a church. Then the light changes, or the mind changes, and any particular likeness becomes more tentative, or more certain: "The poem of the mind in the act of finding/What will suffice," as Wallace Stevens writes in the sentence fragment that begins his poem, "Of Modern Poetry." Magritte's "poetry and mystery," perceived through the necessary temptations of the symbolic, become the power of thought, the action of the mind.

'What will suffice?' is the question that hovers behind everything. But the religious suggestions built into *corpus* don't lead to a particular system of belief. They suggest ways of seeing, acts of finding. "It doesn't represent something else," the artist says. "It means what it is." And now "it" can float away from identification with the work and become the world — it is what it is. In the beginning was the word, and the word became the world, and within the world the work is made that represents the world, and what set it in motion, and what it might mean. Sometimes the light looks like a revelation. Sometimes everything is only itself. Or else: exactly itself.

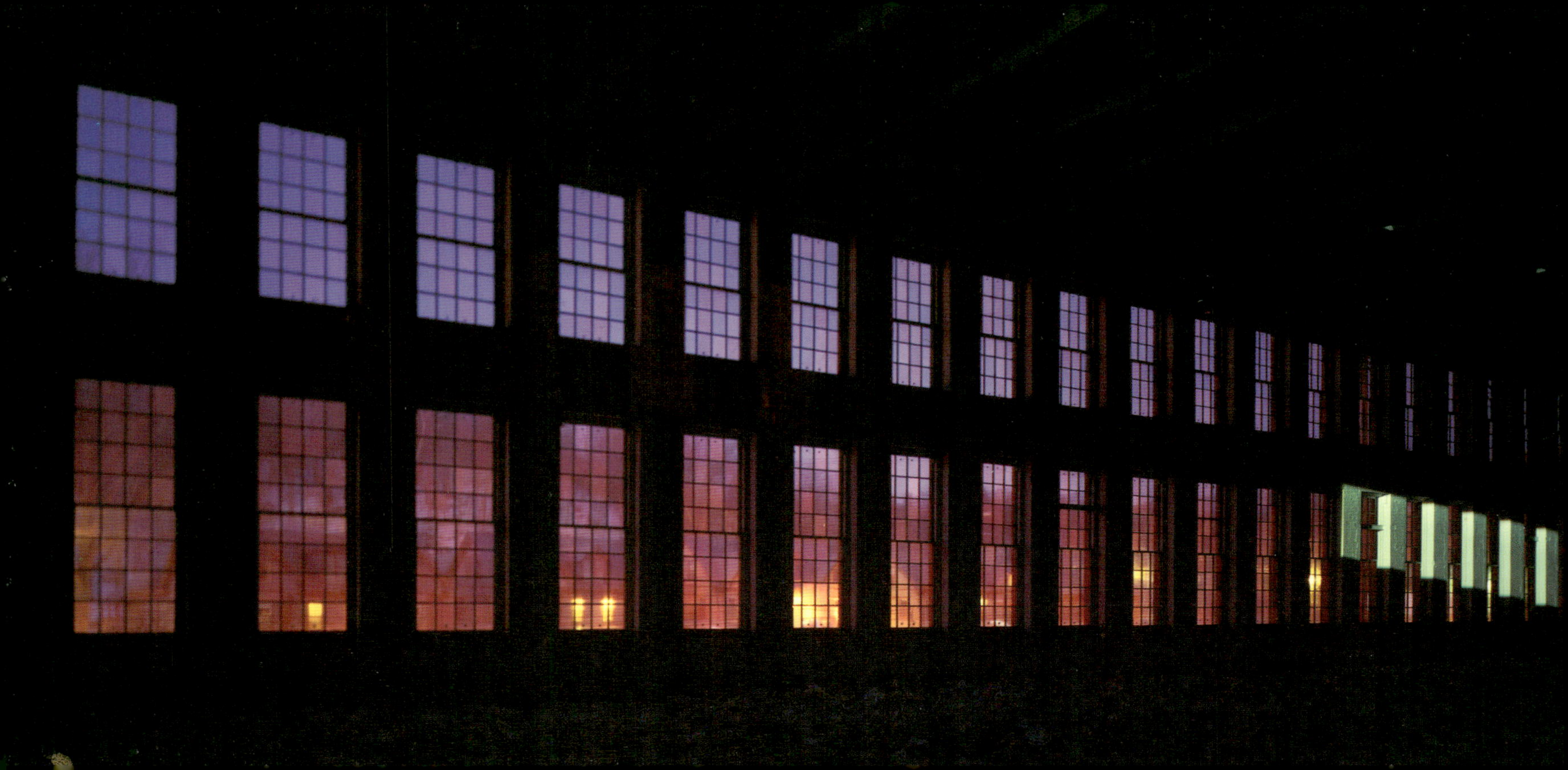

1. Suzi Gablik, *Magritte* (Greenwich, CT: New York Graphic Society, 1970) 11. All but one of the quotations from Magritte in this essay are from Gablik's book, in which quotations are often not precisely attributed. In her notes, Gablik writes, "All statements by Magritte are quoted directly from letters and other texts, either published or unpublished, or from my own record of interviews and conversations with the artist. As with his painting, in his writing Magritte would often work and re-work the same idea. At times his prose is unwieldy and difficult; for the sake of clarity, I have on occasion pieced together relevant statements drawn from different sources to make a composite quote."
2. Ibid., 12.
3. Harry Torczyner, *Magritte: Ideas and Images* (New York: Harry N. Abrams, 1977) 81.
4. René Magritte, "Lifeline" (1940) in Gablik, 183. The complete sentence from which these two phrases are taken is: "The titles of paintings were chosen in such a way as to inspire in the spectator an appropriate mistrust of any mediocre tendency to facile self-assurance."
5. Hugh M. Davies and Lynda Forsha, "A Conversation with Ann Hamilton," in *Ann Hamilton* (La Jolla, CA: Museum of Contemporary Art, San Diego, 1991), 64.
6. Dave Hickey, "In the Shelter of the Word: Ann Hamilton's *tropos*" in *Ann Hamilton: tropos* (New York: Dia Center for the Arts, 1995) 129.
7. Ibid., 129.
8. Charles Bonenti, "Ann Hamilton's Art Installation Speaks in Spaces," *The Berkshire Eagle*, Dec. 12, 2003.

Corpus: MASS MoCA commission
Curator: Laura Heon
Project Managers MASS MoCA: Larry Smallwood, Richard Criddle
Studio Project Manager: Maggie Moore
Engineering: Marty Chafkin, Perfection Electricks
Sound Recording and Engineering: Scott Lehrer, Passport Sound
Additional Engineering: Bill Greenwald
Spinning speaker sound: "liquid air," composed by Meredith Monk for *mercy*, a collaboration with Ann Hamilton. Voices: Theo Bleckmann, Katie Geissinger, Ching Gonzalez, Meredith Monk, Allison Sniffin, Bowed Vibraphone: John Hollenbeck
Unison speaking: Tom Bogdan, vocal leader, with students and staff, Bennington College
Volunteer Interns: Alexis Terry and Olivia Fite. Additional volunteers : Dan Auerbach Althea Bryant, Hyunjung Chae, Karen Combs, Stephan Kelleher, Kyle Mosholder, Carolina Pedraza, Mel Purdy, Bill Ramage, William Ransom, Shaheen Rashid, Meredith Rose, Margaret Smithglass, Rebecca Uchill, Matthew Winkler, Shira Wohlberg

Ann Hamilton is represented by Sean Kelly Gallery, New York

Essay *Acts of Finding*: Lawrence Raab
Photography: Thibault Jeanson; cover, pgs. 9, 10, 12/13, 15, 24/25, 26/27.
Arthur Evans; pgs. 2, 5, 6, 16, 20, 23, 30.
Graphic Design: Arjen Noordeman

MASS MoCA Publications
North Adams, Massachusetts
Printed and bound in the U.S.A. by Mansir-Holden Printing Company.

MASS MoCA

MASS MoCA • 1040 MASS MoCA Way • North Adams, Massachusetts 01247
PH 413-664-4481 • FAX 413-663-8548 • www.massmoca.org

ISBN 0-9700738-9-5